JUN 1 0 2021

NO LONGER PROPERTY OF
SEATTLE PUBLIC LIBRARY

Home Sweet
Houseplant

Home Sweet Houseplant

A Room-by-Room Guide to Plant Decor

BAYLOR CHAPMAN

PHOTOGRAPHS BY AUBRIE PICK

ARTISAN BOOKS | NEW YORK

ALSO BY BAYLOR CHAPMAN

The Plant Recipe Book
Project Succulent
Tabletop Gardens

Copyright © 2019, 2021 by Baylor Chapman
Photographs copyright © 2019, 2021 by Aubrie Pick

All rights reserved. No portion of this book may be reproduced—mechanically, electronically, or by any other means, including photocopying—without written permission of the publisher.

Library of Congress Cataloging-in-Publication Data

Names: Chapman, Baylor, author.
Title: Home Sweet Houseplant / Baylor Chapman.
Description: New York, NY : Artisan, a division of Workman Publishing Co., Inc., [2021] | Includes index.
Identifiers: LCCN 2020042985 | ISBN 9781648290343 (hardcover)
Subjects: LCSH: House plants in interior decoration. | House plants.
Classification: LCC SB419.25 .C455 2021 | DDC 635.9/65—dc23
LC record available at https://lccn.loc.gov/2020042985

Design by Nina Simoneaux
Front cover photograph by Ian Green

Artisan books are available at special discounts when purchased in bulk for premiums and sales promotions as well as for fund-raising or educational use. Special editions or book excerpts also can be created to specification. For details, contact the Special Sales Director at the address below, or send an e-mail to specialmarkets@workman.com.

For speaking engagements, contact speakersbureau@workman.com.

Published by Artisan
A division of Workman Publishing Co., Inc.
225 Varick Street
New York, NY 10014-4381
artisanbooks.com

Artisan is a registered trademark of Workman Publishing Co., Inc.

Published simultaneously in Canada by Thomas Allen & Son, Limited

Printed in China
First printing, March 2021

10 9 8 7 6 5 4 3 2 1

This book has been adapted from *Decorating with Plants* (Artisan, 2019)

*To my parents,
for whom I'll be forever grateful
for encouraging all
of my plant-related pursuits*

CONTENTS

INTRODUCTION

Why bring plants into your home? Because they are amazing. They make the world go round. We eat them, we build houses with them, we wear them and dye our clothes with them, we use them to develop fragrances and to make medicines. Gardening and caring for plants can promote mental and spiritual well-being through a practice called horticultural therapy. Plants are mighty powerful and oh so valuable. Along with furniture, artwork, and decor, plants are a wonderful way to communicate your style and add "soul" to a space. When you get right down to it, plants bring a breath of fresh air to your rooms like nothing else can. Here are just a few things that houseplants can do for you and your home.

STRIKE A MOOD. By bridging the indoors and outdoors and reinforcing our connection to nature, plants encourage us to slow down and settle in. This often happens at a subconscious level: we may not be immediately aware of plants' calming effect, but it's there.

SET A TONE. Delicate flowers communicate a soft, feminine touch, while a structural, dark-hued plant offers an edgier, more modern vibe. An oddball plant, in contrast, will reveal your flair for the curious and cool.

DEFINE A SPACE. Plants can be arranged to direct the eye. They can lead people into or out of a space or create a barrier or "wall" to make one room seem like two.

CLEAN THE AIR. Plants scrub the air in their immediate surroundings by taking in harmful toxins, processing them in their leaves and through the soil, and releasing clean air into the environment. Plants even remove particulate matter such as dust, pollen, and pollution.

MITIGATE SOUND. Plants soak up, buffer, and generally dampen noise. The sizable leaves of the rubber plant, for example, absorb sound waves, while feathery palm fronds will diffract them.

GENERATE AN AROMA. Fragrant blooms and leaves can subtly define a space, soothing or energizing those who dwell in it.

Whether you crave a bit of nature but don't know where to start or you already have a bunch of plants but want to arrange them in a more stylish, composed, or thoughtful way in your space, this book is here to help. Filled with design tips, care info, and a whole lot more, the pages that follow will provide you with the tools and inspiration you need to decorate with plants. Replicate one of the looks featured here, or use the design principles and plant knowledge in these pages to create a style all your own.

Whatever plants you choose, remember these three tenets: First, you don't need to fill your entire house with plants. Just one plant that makes you smile is plenty! Research has shown that even a single flower can improve your mood, and a single plant can significantly clean the air. Second, there's no such thing as a "black thumb." Sure, some people seem to have a natural gift with plants, but everyone can learn how to care for them. Observe your plants carefully, and they'll tell you what they need (you'll learn more about these visual clues in "Caring for Your Plant" beginning on page 20). Tend to them as best you can, and if one doesn't make it, try, try, and try again, until you find one that suits you, your lifestyle, and your home. And third, go ahead and break the

design rules. Scale, color, light, texture, and balance . . . these elements all work together to create a harmonious space, but you needn't be rigid about how you apply them. Move things around until you like how they look. Can't find the perfect spot for a new plant? Give it to a friend! What doesn't work for you may be the finishing touch for someone else.

Plants are magical and can instantly turn any house into a home. You may find that the relationship you develop with your plant companions is actually the most gratifying benefit of all. Some plants will require you to tend to them more than others, but they are worth every bit of time and effort. No matter where you live, you can enjoy a little bit of green.

Nuts and bolts. The foundation. Plants 101. Whatever you'd like to call it, the following pages cover the basic information you'll need to know as you begin to bring plants into your home. Some of these tips will be helpful before you even pick out your first green companion, while other instructions may come in handy months down the road when you're in need of some plant SOS. Feel free to skip ahead to get an idea of what appeals, then come back here for the nitty-gritty. Whether you're ready to grab a plant and get rolling or just looking for an idea of what it will be like to live with plants, this section is here to help.

Choosing and Buying Your Plants

Before you can design with plants, you need the plants themselves. Plant nurseries, boutique plant stores, and garden centers have trained experts on hand to answer your questions and guide you in taking home the right specimen. There are plenty of other places to pick up plants, too, such as grocery outlets, hardware stores, craft shops, and flower stands. If you'd rather shop from home, online plant retailers, as well as sites such as Etsy, offer a ton of options, including hard-to-find rarities (for a few reliable sources, see page 122). Before you head to the store (or the Web), consider these five questions.

1. **Which room do you want the plant to live in, and what is the environment in that space?** What's the light level? Temperature range? Relative humidity? If you have your heart set on a finicky specimen that won't thrive in your space's existing conditions, you can always add "sunlight" with grow lights (see page 54) or create a more humid

microclimate by placing a gravel tray under your pot (see page 25)—but know that maintaining this special mini environment will require more time and effort than choosing a plant that's naturally well suited to the room.

2. **What's your style?** Nature's original decor, plants come in a range of "styles" that can complement the rest of a room's look. Do you want something sleek, or do you crave a wild and untamed mass? Do you like vibrant colors, or would you prefer to stick to a neutral palette of greens? What do you already own that the plant would go nicely with?

3. **How big do you want it to be?** Are you looking for a floor plant or something pint-size for your desk? Ask your local nursery about the growth rate of the plant. The rubber plant, for example, is a rapid grower; if you take one home, be prepared to either offer it significant space to expand or restrict its growth by keeping it in a small planter. Others, like the ZZ plant, will take years to mature.

4. **What's your budget?** Plants, like anything else, vary in price depending on availability, season, location, and growth rate. Choose what works for you and your wallet. The smaller a plant, the cheaper it tends to be. If you are willing to be patient, you might choose a medium-size plant and simply wait for it to grow up into the floor plant of your dreams. (There's another benefit to raising your plant versus buying it full-size straightaway: growing it to maturity in your home results in a stronger plant, one that doesn't undergo the stress of acclimating from its previous environment in a nursery or shop to its new spot in your home.)

5. **How much time do you want to spend caring for your greenery?** Some plants require much more effort than others. If keeping track of when to water, mist, and fertilize makes your head spin, consider relatively low-maintenance houseplants like the sansevieria or the pothos.

WHAT'S IN A NAME?

Every plant has two names: its Latin botanical name and its common name. Think of the botanical name as a bar code for plants—no two are the same. In contrast, plants can have multiple common names, and some common names are shared by multiple plants.

WHAT TO LOOK FOR WHEN BUYING A PLANT

Maybe you've found the exact plant that is on your wish list, or perhaps you've stumbled upon one and decided it just has to go home with you! Before you pull out your wallet, though, pause and take a closer look. To ensure that you're bringing home the best possible specimen, spend a few minutes with your new prospect. Touch it, pick it up, and flip over the leaves. (It's kind of like looking under the hood when you're car shopping.) Here are a few things to check out.

THE LEAVES: If the plant is supposed to be green, make sure the leaves are actually green, not yellow or brown. Are they bushy? Full of life? Avoid plants with wilting or torn leaves, as well as ones with notched or nibbled leaves, both signs of bugs.

THE GROW POT: You want a plant that is proportionally the correct size for its current grow pot—not too large, not too small. The roots should loosely fill the pot. Lift the pot and check the bottom: are roots growing out of the drainage holes? If so, the plant may be undesirably "root-bound." Also called "pot-bound," this is when roots are itching to grow but are tightly constrained to the pot—they keep wrapping and wrapping over themselves to create a tightly knit web of roots. If you were to *gently* lift the plant out of the grow pot, the soil should generally hold together (as shown opposite), not instantly fall apart or come out as a solid mass of roots (though you won't be able to perform this test before buying, unless you ask the garden center staff to help you).

THE BUDS: If flowering plants are what you're after, look for ripe buds, not fully open blooms. This way you can enjoy all the stages of the plant's growth.

Potting Your Plant

Once you get your plant home, you have a couple options. You can "stage" the plant in its original grow pot, or repot it in a vessel of your choosing.

The plant staging supplies pictured here include (clockwise from far left): cork to protect the tabletop, small stones to serve as a decorative topper, a waterproof liner, Red Velvet echeveria, a saucer, a cachepot, and a large stone used to raise the grow pot to the needed height.

STAGING

The cleanest and easiest potting method is to "stage" your plant by keeping it in its grow pot (the original, usually plastic, container) and concealing this unattractive vessel in a cachepot (decorative container). After choosing your cachepot and ensuring that it's large enough to cover the original grow pot, insert a waterproof liner to protect against water leakage or seepage (do this even if your cachepot lacks a drainage hole). Plastic liners are available from florists or garden centers, or you can make your own by using a bowl, the cut-off bottom of a plastic bottle, or even a layer of plastic garbage bags. Make sure this liner is wider than the grow pot and is able to catch any water that drips from the sides of the pot as well as the bottom. If the grow pot sits too far down inside the cachepot, create a "riser" by adding waterproof stuffing (like Bubble Wrap), a layer of gravel, a block of wood, a large stone, or a second, upside-down grow pot to prop the plant up to the ideal height. An added benefit of a riser is that it will act as a protective spacer, keeping the plant's roots from sitting in a pool of water in the event that you accidentally overwater.

MATCHING YOUR PLANT WITH THE RIGHT SOIL

All terrestrial houseplants require a potting mix of some kind—a combination of mediums such as soil, peat, sand, perlite, leaf mold, and bark. Potting mixes are distinguished from outdoor ground soil not only by their light or "fluffy" quality, which provides air gaps that give plant roots access to oxygen, but also by the assured absence of any soilborne maladies.

For the majority of houseplants, regular potting mix is just fine. Some plants, like succulents, may need a cactus mix (one that contains more sand and grit to help water drain) to maintain a drier environment. Talk to your local nursery expert to get advice on a good match.

TOPDRESSING: Every year or two, your plants will benefit from a repotting. For plants that are large or prickly, or that do best with tighter roots, simply remove the top layer of potting mix (and any decorative topping) and add a fresh layer of soil.

REPOTTING

To repot your plant, choose a pot that is roughly the same size or slightly larger than the grow pot (a small plant's root system can't reach all areas of moisture in a big pot of soil, thus leaving the soil soggy and causing the roots to rot). A pot with a drainage hole is best, as it prevents water from pooling at the bottom of the pot and suffocating the roots. Then follow the steps below.

1. Cover the drainage hole with a permeable layer (like a window screen or a coffee filter) to prevent soil from escaping. A layer of gravel is not needed for drainage.

2. If the new pot is larger than your grow pot, add a layer of soil to the bottom of the pot. (Check with your local garden center if you're unsure what potting mix is best for your plant.)

3. Carefully tip the plant on its side, give the grow pot a squeeze, and remove the plant from the pot. Gently massage the roots to separate them (see opposite; this allows them to loosen and grow outside the shape of the original container).

4. Set the plant in its new pot. The base of the stem should rest just below the rim of the pot. Fill in with soil and give the plant a little shimmy to work the soil into all the nooks and crannies. Add more soil if needed. The soil line should rest just below the rim of the pot.

5. Cover the soil with a pretty, decorative topper. Options include moss; rocks, gravel, or pebbles; wood chips; creeping wire vine; and Spanish moss. For something out of the ordinary, try buttons or sea glass. (A topper is like a swept floor—it makes for a more polished look, even if you can't put your finger on why!)

6. Put something between the pot and your chosen surface—a piece of cork, a dish, a pot holder—to add one more layer of protection against water damage.

Caring for Your Plant

Think of your plant as a friend—an exceptional creature in its own right that drinks water, needs nutrients, craves light, breathes air, and even sleeps (or at least rests from time to time). Unlike their wild cousins that receive water, fertilization, and so on naturally, houseplants need our assistance to thrive. Most times when you buy a plant, it will come with a care tag that outlines its light, water, and air temperature requirements. An in-depth look at each of these aspects of plant care, and more, follows.

LIGHT

While plants are adaptable, in order for them to flourish, they should be kept in your best approximation of their preferred lighting conditions. If you set a sun-loving plant in a windowless room, it won't survive for long—but it'll likely do fine with bright light instead of direct sun.

Here are general explanations of the common categories of lighting conditions (essentially, light's intensity drops as the distance from a window increases).

Direct Light

Bright Light

DIRECT LIGHT: Direct light means the sun is shining right on the plant for about half the day.

BRIGHT LIGHT: This is the indirect light you'd get through a sheer curtain or right next to a window that doesn't have sun shining directly onto the surface below it. If you hold your hand over a blank piece of paper in this spot, you should see a defined shadow. A lot of plants like this kind of light.

MODERATE LIGHT: This is the kind of light found a few feet (1 m) inside a room's walls or right along a dim, north-facing window (in the northern hemisphere). There's no exact blueprint for this value because the luminosity depends on many factors, including the size of the window, the orientation of the sun, and any structures (awnings, trees, buildings) that may hamper the sun's rays. The best rule of thumb is that this light will produce a fuzzy shadow.

LOW LIGHT: This is the lighting condition well away from a room's source of illumination (for example, the opposite side of a room with a small window). Note, however, that low light still means some light—you should be able to comfortably read a book where the plant is positioned. Low light is dim, not dark. (If the light is too low, a plant may get spindly or its green leaves will begin to yellow.)

Moderate Light

Low Light

WATERING

When it comes to watering your plant, first consider the type of plant it is and its native environment. A succulent like an echeveria that is accustomed to unreliable rainfall will retain moisture in its leaves and require infrequent watering, whereas a tropical plant like a croton drinks up water quickly and needs ambient humidity.

Watering frequency then will depend on several additional factors: the season (winter offers shorter days and cooler temperatures, meaning less water is needed) and the climate (Florida is more humid than Colorado), as well as the pot placement (a stand-alone planting in a sunny spot requires more water than a grouping in the shade), pot material (a breathable clay pot dries out more quickly than a plastic one), and pot size (a tiny pot needs to be watered more often than a large one). Two of the same exact plants, if placed in different environments, would require different watering schedules.

As a rule, water your plant when the first inch (2.5 cm) or so of soil is dry. Use your finger, a water gauge, or a chopstick to measure soil moisture (for the latter, a brown line will appear on the chopstick at the point where the soil is moist). Weight is also a helpful indicator—lift the pot before and after a good soaking to feel the difference between parched and drenched soil. Beyond that, watch the plant—it will likely wilt when it needs water. If the plant hasn't been without water for too long, it'll bounce right back once you give it a drink (it's cool to watch the leaves' response as they perk right up and say thank you!). Following is a rough guide on the levels of moisture your plants may require (and remember that plants may need different levels of moisture at various times of the year).

DRY: If your plant requires dry soil, that means times of drought are okay or even beneficial. Let it dry out completely for a time before watering thoroughly. Plants benefiting from this watering method, or lack thereof, often have plump leaves, like succulents, or have adapted other clever ways to store water in their naturally drier climates.

SLIGHTLY MOIST: This level of moisture works for the majority of houseplants. The key is to ensure that the top 1 inch (2.5 cm) or so of the soil dries out before you water again (for a small 4-inch/10 cm pot, it'll be the first 1/4 inch/6 mm).

EVENLY MOIST: For plants that require evenly moist soil, don't allow the soil to dry out at all between waterings. Moist soil is comparable to a wrung-out sponge—soft, plump, and slightly wet, but not soggy.

How to Water Your Plant

There are many ways to water a plant: soak it in a bowl of water, the sink, or the bathtub and let it drink from below (don't soak for longer than an hour); sprinkle water from above onto the leaves; precisely water the soil below the foliage; mist the leaves; or let it make its own "rain forest" in a cloche or a closed terrarium (see page 25). Some plants prefer certain methods, but when in doubt, the best option is to water the soil, not the leaves. Whatever method you choose, room-temperature water is best. Generally speaking, when it's time to water, let the water fully saturate the soil. After watering, place the vessel on a saucer, let the water run out of the bottom of the pot, and discard the excess water.

Watering FAQs

Below are a few common questions you may have when it comes to watering your plant. If your plant just doesn't seem happy, or isn't drinking up its water like it should, consult this rundown for a few easy fixes that'll perk a thirsty plant right up.

Is the water running off instead of soaking in? If so, the soil may be compacted. Make a few holes in the soil with a chopstick or fork and water again. The plant may also be root-bound (see "The grow pot," page 15); in this case, it's time to repot to create more space for soil and moisture to mingle in with the roots.

Has the soil separated from the edges of the pot, or is water quickly running right through the pot and not soaking in? This is an indication that your soil has completely dried out and needs to be rehydrated. If your pot has drainage holes, hold the pot in a bowl of room-temperature water. Let go when it starts to absorb water and stops floating, then soak for an hour or so to let the soil fully absorb moisture. If your pot doesn't have drainage, fill the vessel with water, wait, and repeat until the soil is plump. Then gently tip the vessel out over a sink to drain any pooling excess water.

Think you've overwatered? Gently free the plant from the pot (it should come out with soil and roots intact). Then lightly pat down the edges of the soil using paper towels to absorb some of the excess water. Let the plant stand, unpotted, on a plate

Use a narrow-mouthed watering can to more precisely water plants that don't like to get their leaves wet, like this Cape Primose.

until the soil becomes just moist to the touch, then return the plant to the pot.

Going on vacation? If you'll be gone for only a week or so, a simple solution is to give your plants a good drink before you leave and cluster them together to raise humidity and slow evaporation from the soil. You can create a similar effect by placing pots on a gravel tray (see below) or even gently covering your plants with lightweight plastic (think dry cleaner bags) to create a temporary greenhouse effect. For a longer trip, consider purchasing a water globe, a wick-watering mechanism, or even self-watering pots. You can also make your own water-releasing contraption: Set a jug of water next to your plant. Cut pieces of natural twine into sections long enough to reach from the jug to the plant, then place one end of each piece at the bottom of the water jug and stick the other an inch (2.5 cm) into the soil surface. The standing water next to the plant will increase humidity, while the twine will replenish moisture in the soil as it dries out. Of course, you can also ask a friend or neighbor for a favor or even hire a professional plant sitter!

HUMIDITY

A lot of plants (and humans) feel most comfortable in 60 percent relative humidity. If you have artificial heating or cooling in your home or live in an arid climate and you are growing humidity-loving tropicals, you'll probably need to increase the moisture in your air. You can achieve this in myriad ways: grouping plants together; covering or containing your plants under a cloche or in a terrarium; adding a humidifier to the room; misting your plants; moving them to a more humid room, like the bathroom; or setting your pots on a waterproof tray filled with gravel and adding a thin layer of water (make sure the gravel keeps the pots above the waterline).

TEMPERATURE

Plants thrive with fresh air and good air circulation. A gentle, warm breeze from an open window on a lovely day is a welcome balm for most. Be mindful of blazing radiators, arctic air conditioners, and drafty winter windows, all of which create conditions too extreme for most houseplants. Some variation in temperature is okay, however. Just like their wild cousins, houseplants can benefit from cooler evening temperatures. And although plants are sensitive to temperature, they are adaptable. The recommendations below aren't strict rules—they're ideals.

CHILLY: Below 55°F (13°C) but above freezing. This climate is detrimental to most houseplants, but there are a few that benefit from it, including bulbs that require forcing.

COOL: Roughly 55°F to 60°F (13°C to 16°C). For reference, this temperature may give you a slight chill, like when you enter a garage or a basement.

AVERAGE: 60°F to 75°F (16°C to 24°C). This is a wide range, and one most houseplants (and humans!) will feel fairly comfortable in.

WARM: 75°F to 80°F (24°C to 27°C). This is the climate for plants that like things toasty.

Keep light and temperature conditions in mind when choosing plants for in front of a window. Jasmine (in the birdcage) and Thimble Cactus (on the windowsill, right), for example, thrive in direct light and are hardy enough to withstand occasional fluctuations in temperature.

FERTILIZER

Like us, plants require nutrition. Some nutrients are obtained from the potting mix your plants are grown in, but you may want to supplement with a fertilizer during the longer days of spring and summer. The majority of houseplants will benefit from a monthly application, but some plants require a more regular regimen (twice monthly) and others will benefit from a lighter application (every two or three months); it's often best to cut out fertilizer completely in the winter months (when growth tends to slow or the plant enters a dormant period and needs rest).

Fertilizer is offered in powder, liquid, and pellet form, and will be marked with three numbers that indicate the percentages of nitrogen (N), phosphorus (P), and potassium (K) in the mixture. Different combinations promote leaf growth, root growth, blooming, and fruiting, but a "general-purpose" natural fertilizer offers a solid one-stop-shopping option. Natural blends, such as fish or seaweed, are also a great option that offer more wiggle room in terms of timing and measurements, since they generally supply lower levels of nutrients than chemical fertilizers do. If opting for a liquid fertilizer, water the soil first to increase absorption.

Regardless of the formula you choose, it pays to be conservative in your application—less is more. Overloading the soil with nutrients is often more detrimental than avoiding fertilization completely, as it can burn the plant's roots and cause distress. Especially with chemical fertilizers, skip the manufacturer's instructions and use a half-strength mix instead.

Plant SOS

Plants show you what they need. Caring for them is a matter of understanding and interpreting their signals and, when needed, changing things to keep them healthy. Sadly, a lot of plants are killed with kindness—often with too much water. To make the situation worse, after just one bad run, folks say, "I'm a plant killer!" Having a plant die on your watch may be disheartening, but please don't give up. It's all about practice, trial, and error. Here are some common maladies and possible solutions.

WILTING OR CURLING LEAVES: The most common cause of wilting or curling leaves is either underwatering or overwatering. If the pot feels lighter than usual, it is a good indication that the soil has dried out and needs water. Conversely, if when you lift the pot there's a pool of water in the saucer, you've been overzealous with the watering can. Empty the saucer and soak up any sogginess in the soil with a rag or newspaper. If water doesn't seem to be the issue, consider temperature. Heat can make a plant wilt, and curling leaves can be brought about by cold drafts near windows, doors, or open hallways. Aphid bugs can also cause curling, as they suck the juices out of a plant's leaves. If you spot these soft-bodied bugs, wipe them away or blast them with a faucet's spray to remove them.

BROWN OR YELLOW SPOTS ON LEAVES: Spots and patches on leaves are usually a sign of either a watering issue or pests. Do some investigating—brown, crispy tips signal underwatering or low humidity; brown soft spots may mean you've been overwatering. It could also be a fungus or blight: remove any damaged leaves and make sure your plant has good air circulation. Speckles on the surface of the leaves or a blister-like appearance may indicate that you have unwelcome scale insects. Remove these critters with a cotton swab soaked in rubbing alcohol, and be diligent—they're tough to conquer once they've interloped. Finally, if you've recently moved your plant from shade to a sunny spot, it may be sunburned; always move plants from shade to sun in stages to acclimate them.

RAGGED OR HOLEY LEAVES: If holes or notched edges suddenly appear on the leaves of your plant, it's likely that the plant is hosting some sort of chewing insect. This can be treated fairly easily—pick off any adults and spray the affected area with soapy water. Evidence of snail activity usually includes a trail of shiny slime. Look in nooks and crannies where snails might be hiding from daylight.

WRINKLED OR SHRUNKEN LEAVES: Naturally fleshy plants like succulents (including cacti and sansevieria) will let you know when they're receiving too much or too little water. Gently touch a leaf: if it's hard and wrinkled, it's time to water; if it's soft and prune-like or has gone to mush, it's a sign you may have been overwatering.

FALLING LEAVES: It's quite normal for leaves to fall off some new or recently repotted plants. Give those plants time to acclimate and prevent shock by slowly and gradually moving them from an area of low light to one of bright light. If you haven't recently moved or repotted your plant, falling leaves could be due to an undesirable temperature change or could signal overwatering or underwatering (check your plant's requirements). If the leaves progress from healthy green to pale then darker yellow and finally fall off, give your plant a gentle shake—if a cloud of tiny flying bugs appears, you're dealing with a whitefly infestation (for treatment options, see Bugs, opposite). And remember, it's natural for leaves to fall off your plant—old leaves will sometimes drop and be replaced by new growth.

LEGGY STEMS: Spindly growth, also called etiolation, is a hint that your plant isn't receiving enough light and is on the prowl to find it. The flat rosettes of an echeveria are particularly susceptible to such wanderings if left without adequate light. (Other telltale signs your plant is light deprived include small leaves, pale green growth, and leaf loss.)

STICKY STUFF: Leaves that are sticky to the touch may be home to unwanted guests. The stickiness is not from the plant but is a substance called honeydew, excreted by an insect pest. There are many different pest treatments out there. Most are pest specific, so it may require a process of elimination to find the right remedy.

GUTTATION: This is not as horrific as it sounds! Guttation is the term for tiny water droplets forming on leaf tips. Unlike dew, which settles on the leaves, this water comes from *inside* the leaf and is caused by wet soil (when the plant takes in too much water and pressure builds, which forces moisture to be exuded as little droplets). You're most likely to notice this phenomenon in the morning before moisture has had a chance to evaporate.

ROOTS ABOUND: If roots are popping out of the pot's drainage holes, it might be time to repot. Gently lift out the plant; if it's one mass of tightly knit roots, it's time for an upgrade. If you want your plant to grow larger, slightly increase the pot size: break up the roots with your hands (or a clean knife), remove a portion of the soil, add some fresh potting mix to

the pot, and replant. (See page 19 for more on repotting.)

BUGS: The first layer of defense against pests is to keep your plant healthy. Insect troublemakers are on the lookout for an easy prospect and are more likely to infest a run-down plant than a vigorous one. If you find an insect intruder, segregate your plant (or even set it outside) until it has healed, and be consistent with your treatment. If you catch the bugs early (and they aren't flying off everywhere), start by simply washing them off with water. You'll need to rinse, wipe, and repeat for a few days after they disappear. For a stronger line of defense, try applying rubbing alcohol to the infested areas with a cotton swab, or mixing a few drops of natural dish soap with water and applying with a spray bottle or cloth. For specific treatments, you can take a photo of the infestation and bring it to a plant nursery, or Google "master gardener" to find your local expert and ask for assistance. If things get serious, and bugs abound, consider escorting your plant to the compost bin so that those creepy crawlers don't spread to other plants.

Shaping Your Plant

Plants have their own natural forms (draping, upright, bushy), but with help from humans, they can be manipulated into all sorts of shapes. Bonsai plants, for example, are kept small, while topiaries are groomed to a particular form, and espaliered plants are trained against a wall. Stems can even be grown over structures to create curved configurations or woven into living art (see the braided spear sansevieria on page 112, for example). Here are a few simple techniques for shaping your plants.

PRUNING

We prune plants both for aesthetic reasons and for the health of the plant. You may want to keep your plant a certain size or make a straggly or spindly plant fuller and bushier by trimming to encourage branching and growth.

Plants can also be pruned into whimsical shapes like elephants or more classic topiary shapes like lollipops. The variegated dwarf umbrella plants on the opposite page, for example, have been pruned and trained differently from the start of their lives. If allowed to grow naturally, the plant on the right will continue to spread up and out in a wild bushy form. In contrast, the plant on the left has been carefully pruned, with new growth removed along the trunk, to create its tree-like shape. If a plant has a broken stem or a section ravaged by diseases or bugs, pruning will freshen it up. And once a flowering plant is done blooming, deadheading (snipping the spent blooms) will keep it from setting seed and thus may encourage it to bloom again.

ADDING SUPPORT

The first form of support you may encounter is a stake tied to the main stem or trunk of your plant when you purchase it from a nursery. This holds the tree or

Pictured here are two different variegated dwarf umbrella plants (Schefflera arboricola 'Variegata'), one pruned and trained (left) and the other allowed to grow wild (right).

plant upright during transportation. However, the green tape and bamboo poles often used to secure plants might not be aesthetically pleasing, and the overall shape of your plant may need some adjusting. To restake your plant, first carefully snip the tape that holds the plant's stems to the stake. Move from the bottom to the top, slowly supporting each stem as you go with your hands. Gently let the stem go. Repeat with each stake. Once the stems are free, shape and support as you wish. Experiment with green twine and stakes that can be camouflaged, or go for something bright and attention grabbing.

Support structures can take many other forms, too: a hidden stake around which young plant stems are braided (as shown here, on the left), a series of wall hooks supporting a vine (see page 88), or a trellis on which you've espaliered a plant (see page 94).

Finally, pruning and adding support go hand in hand. You'll need snips and twine or twist ties handy to help guide and support your upwardly or outwardly crawling green friend.

Sharing Your Plant

Sharing a plant is a meaningful endeavor. It's an easy and affordable gesture of trust, friendship, and community spirit. Plus, seeing how someone else shows off your living gift—with vase choice, placement, planting technique—is a fun way to collaborate, learn, and be inspired. Here are a few simple ways to share your plants.

OFFSETS: Offsets are essentially baby plants that are produced from a mother plant. With plants that produce offsets, like bromeliads and some succulents, simply pluck or cut, and share!

ROOTING: Rooting is the process by which a cutting or piece of plant is placed in water, perlite, or a potting mix until new roots form. Adding rooting hormones can help with faster and better root formation. Pothos are good candidates for this method. Once roots form, the tiny plant can be planted in a potting mix.

DIVIDING: Sometimes you must dig deep into the soil to propagate plants. To divide a plant, unpot your specimen and separate the root clump. Once separated, replant in fresh potting mix.

LEFT: *Dividing plants can take some muscle, especially when the roots are tenacious (like in the case of this ZZ plant). Don't be afraid to really yank them apart or, if needed, employ a sharp knife for assistance.*

CHILD AND PET SAFETY:
In the context of this book, *toxic* does not automatically mean deadly—severity and symptoms vary. As a precaution, you may want to keep all plants out of reach of small children and animals. Some plants, like the ZZ, aren't dangerous to touch but are harmful if ingested, while others that deter predators with sharp thorns (like cacti) or irritating sap (some euphorbias) may injure your skin with the slightest touch. Visit the ASPCA website (ASPCA .org) for a list of pet-safe plants.

Once you've picked out a plant and learned its care requirements, it's time for the fun part: figuring out how to style it! Decorating your home with plants ought to be fun and satisfying, not intimidating. The only "rules" you need to follow are the care guidelines of your chosen plant (see pages 20–28). Other than that, experiment until you find a look that's right for you! The information that follows offers some tips and tricks to help get you started.

Plant Attributes

Every plant you bring home has a personality all its own. The first step in deciding how best to style it in your space is to take some time to really look it over and get to know it. Here are a few important features to consider.

SIZE: Though houseplants come in a wide range of sizes, from teeny-tiny succulents to towering palms, for the purposes of this book, plants are described as either floor plants or tabletop plants. Floor plants are your statement makers. These bold specimens can single-handedly change the feel of a room. A tabletop plant, on the other hand, is just the thing for smaller spaces or to use as an accent piece or part of a more layered look.

COLOR: When you envision a plant, you might conjure up just one shade: green. But plants are a colorful bunch! From silver to copper, canary yellow to sunset pink, plants come in colors across the spectrum, and even those that are all green tend to boast numerous verdant tones within.

TEXTURE & PATTERN: Furry, glossy, polka-dotted, or rippled: your plant's textures and patterns are the features light will bounce off and the elements bound to turn heads. Humans are tactile creatures, after all. Think about all aspects of your plant—both sides of the leaves, and even its bark and trunk—and make sure you position it to show off your plant's full outfit (see more on placement on page 40).

FORM: Plants comprise many types of shapes and structures. Each form communicates a certain feeling and mood. In this arrangement at right, an arching Buddha's hand elephant ear (*Alocasia cucullata*, top) has an airy, carefree presence (imagine it waving in good luck around Buddhist temples in tropical areas of Asia), and a draping spider plant (*Chlorophytum comosum*, middle) lends a feeling of softness and whimsy, particularly here with this 'Bonnie' curly variety. While each plant is unique on its own, a combination of different forms can create a complementary grouping, like with the three shown here. The bold starfish sansevieria (*Sansevieria cylindrica* 'Boncel', bottom) signals strength and adds a sense of structure to the whole ensemble.

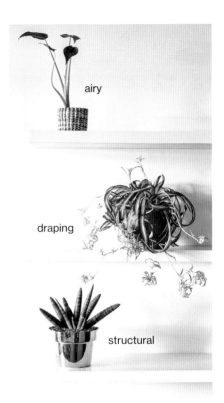

airy

draping

structural

Vessels

In the pages that follow, you'll see many types of vessels, from classic terra-cotta pots to hip brass cachepots to repurposed soap dishes. Don't be constrained by the notion that plants belong in common containers. Look around the house— almost any vessel (or bag or box) can be turned into a cachepot to conceal a plant's original plastic grow pot.

Your goal is to pair plant and vessel in such a way that they will play off each other, elevating the whole look. Here are a few factors to consider.

COLOR: The right color container can make your houseplant sing, while another hue will make it look subdued. For example, a green leafy plant will pop against a contrasting red vase, whereas its color would recede against a matching green one. Consider the room as well. What colors do you want to surround yourself with? If stuck, the matte clay of an easy-to-find standard terra-cotta pot makes the leaves of almost any plant radiate.

SHAPE: The shape of vessel you choose will depend on the architecture of the plant. The classic flowerpot shape (where the opening tapers down to a narrower base) is like the little black dress for plants: it makes almost any variety look good.

A pot that is the same width from the opening to the base may make some plants look stocky—use this shape if your chosen greenery is tall or wide enough to balance it out. An hourglass-shape or pedestaled vase works particularly well for a draping specimen.

Shape also communicates a mood—a square vessel telegraphs stability and firmness, while a round one will feel approachable and warm.

STYLE: Finally, consider not only the plant's personality but also yours. If your home has a natural, rustic feel, you might want to stick your fern in a log vase, but if you're flashier, opt for a neon bowl. Whatever vessel speaks to you is a fabulous choice!

ONE PLANT, TWO WAYS

To highlight how impactful your choice of vessel can be, two different vessels are used below. As discussed on the facing page, color, shape, and style all impact the overall appearance of a plant-vessel pairing. Here, the focus is on the power of shape, keeping the color and texture of both vases neutral.

THE HEIGHT OF ELEGANCE

A tall vase is a fitting match for any plants that drape. After a riser and a pot liner were placed in this terra-cotta pot's cavity, in went the plant in its original grow pot. The vase's curved outline adds grace and femininity to the look, and its height emphasizes the dangling flowering stems as they delicately spill out of the vessel. The warm, natural tones of the terra-cotta lend the pairing a classic, garden-like feel.

A JUST-SO SPRAWL

The clean white veneer and widemouthed opening of this salad bowl make the unexpected vessel a pleasingly minimalistic partner for this funky chenille plant. The lush vine-like stems gently drape over the bowl's edge, just grazing the surface below, resulting in a naturalistic composition. (Be careful when pairing draping plants with low planters: if this bowl's edges were much lower, the stems could have ended up looking floppy and lifeless.)

Design Basics

The pages that follow are filled with tips to get you in the groove when it comes to designing with plants, but below is an overview of a few key concepts.

SCALE: To determine if your plant is properly to scale, consider it in relation to three things:

1. **Its vessel.** As a general guideline, planters should be roughly one-third the height and/or width of your plant. For a more "Zen" design, try the opposite balance.

2. **The furniture around it.** A chunky couch will be able to stand up to denser planting better than a slender chair.

3. **The rest of the room.** Is the plant touching the ceiling or getting cramped in a corner? If so, it might be too big for the space. Conversely, is it getting lost among the large statement pieces in the room? Maybe it needs a new location, or more plants to surround it and lend it visual mass.

BALANCE: Balance is the key to a pleasing design. In a basic sense, balance is the even distribution of weight. When you're talking about this in terms of home design, the "weight" you are looking to distribute is the visual interest in the space. Generally speaking, you can do this one of two ways: by creating a symmetrical arrangement or by creating an asymmetrical one. Symmetry is easier to achieve but asymmetry offers more flexibility to play with your design. See pages 58-61 for two living rooms that deconstruct these approaches.

PLACEMENT: There are many places to display your plant, whether hung from the ceiling, set on a pedestal, or resting on a shelf. The shape of both your plant and its vessel will determine much here: if it is a draping plant, let it drape—hang it or set it atop a pedestal; if it's a climber, give it something to climb.

Scale also plays a role. A huge pot set on a narrow shelf would look like a boulder waiting to fall on your head. Conversely, a tiny pot on the floor may look like it has been abandoned! And think about how you'll interact with your plant and the distance you'll be from it. You'll be able to enjoy far more detail on a plant that shares your desk space than on one you walk by in the hallway.

VANTAGE POINT: Consider the view you'll be gaining when situating your plant, too. Some plants are best seen from directly above—place these at lower heights to get the most bang for your buck. Others flaunt their stuff when viewed from below, so put these high up on shelves to enjoy their full potential.

Low windowsills are a great spot to highlight tiny plants with dazzling patterns or shapes, like this collection featuring (from top to bottom) a haworthia, ivy, spear sansevieria, 'Red Ripple' peperomia, and ZZ plant.

Entryway

The entryway provides a buffer between the outside world and your living space. You step in and take a deep breath, and a switch is flipped—you know you're home. With a little time and effort, you can use plants to create a smooth and pleasant transition from outdoors to in. You'll find solutions here for any kind of space, from a crowded hallway that needs to accommodate a bustling family to a grand foyer aimed at impressing guests. Strategically placed greenery can direct visitors' views or pique their curiosity by partially screening what's around the corner. Well-chosen fragrant and colorful plants in particular can also be used to set a tone. Whatever sort of space you have and whichever plants you choose, the most important thing is to let your personality shine—all the better to signal "I'm home."

A GREEN WALL IN YOUR ENTRYWAY

The entryway is a hub of activity—boots get kicked off, coats dropped, groceries set down, and homework forgotten. But there's always room for plants! The key for bustling entries is to keep plants out of the way, whether in a basket hung securely on the wall, on a shelf, or in a custom wall unit. This will make it less likely that people will topple the plants in the whirlwind of getting out the door. Plus, hanging plants at eye level turns them into a full-on display.

The system shown opposite, created with a combination of wall-mounted pieces found at Pottery Barn, has a cohesive overall look thanks to a unifying color palette of blues, browns, and pinks. Though the overall effect is one of tidiness and order, the oversize 'Painted Lady' philodendron and the moth orchid (whose stem extends beyond the backdrop of the brown shelf) bring a natural softness to the look. The China doll plant (to the right of the moth orchid) is rather plain, but it pops off the shelf thanks to its white vessel with a green pattern. The leather wall pocket piece was sold as a garden-shed organization tool, but it works nicely as a mail sorter and air plant holder. Bonus: The *Tillandsia caliginosa* air plant (bottom right pocket) is wonderfully fragrant when in bloom.

LEFT: *Did you guess that this petite pink plant is a poinsettia? Poinsettias are euphorbias and come in many shapes, colors, and sizes—see page 121 for another type. This one was placed in a felt basket (by Swedish company Aveva) lined with a coating to create a waterproof, nonbreakable vase.*

VERTICAL GARDEN

If your entryway is a true pass-through, hang a living wall to keep plants in sight but out of the way and save all surface areas—including the floor—for other uses. Re-create the design pictured here using simple shelving brackets and long metal planters with these seven steps.

1. Choose the vessels. These low, long metal troughs can be found at flower shops or home-goods stores, but any lightweight vessels with flat backs will do. These were already lined with plastic, but if yours aren't, waterproof them by lining them with cellophane and/or individual plastic liners.

2. Select the plants. Mix textures, add pops of bright yet related colors, and include a welcoming burst of fragrance. Avoid plants with suckers, such as creeping ficus, ivy, philodendrons, and similar plants if you don't want them climbing up the wall and potentially causing the paint to chip. Here 'Sharry Baby' oncidium orchids, guzmania bromeliads, and jasmine grab the attention, with a combination of fine- and wide-leaved ferns as the backdrop.

3. Design the plantings. Keep the majority of the plants the same from container to container for visual continuity. Let a couple of the containers act as supporting players, and keep them simple so that the others can shine for a more compelling design. Make sure the plants' grow pots aren't taller or wider than the vessel (cut off any offending plastic if one is slightly too tall).

4. To visualize where the planters will hang, measure each planter and create actual-size cutouts with scrap paper. Attach the cutouts to the wall with painter's tape, then use the tape to vertically "draw" in a few plants' heights. Step back and take a look. Stand at the front door and take another look. Move the cutouts around until your design is pleasing. For a

lush look, hang the vases close to one another and fill the wall with green. Or go for a sparer design with just one or two vessels.

5. Hang each vase with a bracket (ideally, a French cleat and a bottom spacer that allow for a plumb hang and ample airflow between the wall and the planter to prevent dampness and mold). It's best if your living wall is easily detachable so you can clean away dust and bugs that linger and hide behind the installation.

6. Set each plant inside the vessel in its original plastic grow pot. If the grow pot is too short, prop it up with something waterproof like Bubble Wrap or an upside-down plastic cup.

7. Follow the care directions for each plant. If ease of care is important to you, be sure to choose plants with similar water and light requirements.

ZEN ENTRY

To turn your entry into a quiet retreat from a busy world, keep your planting simple. Choose soft shapes, a harmonious and subdued color palette, and an unobtrusive container, and be sure to maintain an open path that allows for passing by with ease. This design takes its inspiration from the Saihoji Kokedera (moss temple) garden in Kyoto, which is mostly swaths of green mosses and where visitors instantly feel relaxed upon entering through the garden gates. To mimic that calming experience at home, display a prepotted bonsai tree, or re-create this mini Japanese landscape with a handful of easy-to-find plants.

WHAT YOU'LL NEED:

Ferns and mosses (pictured on page 51: cretan brake fern, Sprengeri fern, spike moss, sheet moss, and mood moss)

Wide, shallow bowl

Potting mix

Smooth, protruding rock

1. Set the plants inside the bowl to plan the layout of the arrangement. Although it is the combination of all the plants that makes the design beautiful, the space that holds the subtle moss ripples is the most impactful feature. Let the tranquil power of the negative space radiate; be thoughtful and add other plants sparingly.

2. Soak the moss in water.

3. Fill the bowl about three-quarters of the way with potting mix—leave enough room so the plants and moss will rest below the rim when they're added.

(continued)

4. Add a few more scoops of soil and, with cupped hands, mold the soil piles to form tiny mounds.

5. Place the rock off-center in the bowl. Unpot and plant a small fern, tucking it in at the base of the rock as if it grew there naturally.

6. Plant the biggest fern at an angle, letting it drape over the edge of the bowl.

7. Grab clumps of the soaking moss and give each a squeeze to release the water. Gently rest the moss on the soil—mood moss naturally mounds and creates an undulating layer of green, but the mounds give any type of moss a lift. Tuck in a piece of sheet moss or any other kind of moss in the fall of the ripples to create a wave-like moss meadow.

8. To care for the arrangement, keep a spray bottle handy and mist daily with water. Bright or moderate light works well for this planting but keep it out of direct sun.

A LOW-MAINTENANCE CONSOLE

If a simple-yet-strong style statement with easy-peasy plants is what you're after, 'Janet Craig' compacta dracaena, starfish sansevieria, and 'Fernwood' sansevieria (pictured opposite, on the bottom shelf) are calling your name. These plants tolerate low-light conditions and won't require frequent watering. To add fragrance and a pop of color, pick up a zygopetalum orchid (top left). These prefer a bit of light (though no direct sun) and weekly watering while in bloom. See below for two more winning options.

A WARM WELCOME

Pineapples were once a luxurious rarity grown, displayed, and eaten only by the wealthy, who shared them with special guests. With such a history of hospitality, a potted pineapple plant makes for a fitting addition to your entry. Pineapples need light and moist soil. After the fruit ripens to a golden yellow, cut it off and slice it open for a sweet treat!

FEMININE FLORALS

If you're craving the beauty of a cut floral arrangement but want the ease and enduring quality of a living plant, opt for a blossoming specimen from a greenhouse plant nursery. Here the contrast of the fluffy rosettes of the tuberous begonia and the dark veiny backs of its leaves are oh-so gorgeous.

SHED SOME LIGHT

If you're an apartment dweller with a front door that leads to an interior hallway, or you otherwise have a very low- or no-light entry, consider hanging grow lights. The cheery contraption shown here is a hanging terrarium with programmable LED to mimic the sun, purchased from Modern Sprout's online shop (while incandescent bulbs emit only the yellow and orange portions of the spectrum, LEDs produce the full spectrum of light plants need to survive). It gives the impression of a bright window in a windowless room, subliminally connecting viewers to the outdoors, even though it's all a mirage.

To create a similar setup yourself, pick up some full-spectrum LED grow strip lighting (narrow rolls of LED that can be rolled out and adjusted for easy adhesive-strip mounting) and attach the strip (light side downward) to the inside top of a shadow box, upcycled vintage fruit boxes, or any shelving unit.

RIGHT: *If you purchase a programmable LED unit, as shown here, set a timer for the electric sunshine: Most plants are happy with about twelve hours of light a day. See pages 20–21 for tips on how to know if your plant is getting the right amount of sunlight.*

Living Room

The living room is one of the most public places in the home and the space where our outward selves reign— the room and its decor are an expression of who we are and what's important to us, plants included. A topiaried tree in a formal living room, for example, might suggest that you like things a bit more controlled, refined, and cared for. The same specimen left to grow wild and woolly might say that you're laid-back and love nature for what it is.

Any size plant is worthy of showing off, but since the living room is often the largest room in the house and tends to have ample floor space, this chapter features lots of big plants, including floor plants and small trees, with tips on how to layer them with a fabulous supporting cast of medium-size green companions.

STATEMENT-MAKING SYMMETRY

In this living room, a few well-chosen plants add life and warmth to a bold, clean design. A formal, symmetrical pairing of towering foxtail palms in matching industrial planters flanks the couch, with a coffee table (topped with a nerve plant) centered between the planters.

There's more to a well-balanced space than the physical arrangements of the objects in the room. You want to balance the *feeling* of the space as well. Though the size and placement of the foxtail palms telegraph strength and power, their curved stems frame the couch to create a canopy that softens the look and adds coziness to the room. Similarly, the dark colors of the couch and metal planters are offset with lightness: elevating the plants provides negative space and lifts the "weight" off the floor. Finally, a repeated triangular pattern woven through the room's design (in the lattice of the metal coffee table, the artwork, the leaf shape of the foxtails, and the knitted throw) ties the whole look together.

Floor plants should reach up to, but not touch, the ceiling. A tall plant resting several feet below the ceiling will draw the eye down rather than up, making the space feel cavernous instead of generous. If your plant isn't tall enough on its own, you can add a plant stand (as we did here).

LAYERED ASYMMETRY

The beauty of an asymmetrical look like this one is the way a viewer's eye moves gently across the scene, from the oversize fiddle-leaf fig on the left to the intricately patterned specimens under the coffee table to the yucca peeking out from behind the couch. There's a rhythm to it all—and a balance. The arrangement is anchored by the center of the couch, coffee table, and set of art prints. From that foundation, the height of the fig on the left is offset by the relative mass of the plants on the right. More greenery and decor were then easily added to this well-balanced design. For a layered design, tuck plants into unexpected places. Here are a few tips.

CREATE A BACKDROP. Use the space behind a sofa to display floor plants or plants set on a table or plant stand.

ADD INTEREST CLOSE TO THE GROUND. Place a plant next to or slightly under a table, for example, or at the corner of a bookcase or couch.

THINK BEYOND THE PLANT STAND. Create a tower of books and use it to lift a plant to just the right height.

STANDOUTS FOR THE COFFEE TABLE

The coffee table is a key complement to your couch, perfect for resting teacups or your stocking feet on. This is a high-traffic space, and real estate on the table is precious. When adding a plant, it's best to keep things simple. Here are a few options.

KEEP IT LOW

For a low-profile plant, place a draping specimen like this 'Hope' peperomia in a large, shallow bowl. Let its waxy leaves gently tumble over the edge of the container and table. Here the peperomia's root-ball is wrapped in moss, a practice called kokedama.

STACK 'EM UP

Succulents like this Devotion echeveria are often planted in simple terra-cotta pots. Rather than tucking it into a cachepot, simply stack the terra-cotta container on top of other decorative bowls. This quirky styling trick prevents any excess water or moisture from damaging the table.

SHADOWY SHOW-OFFS

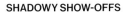

In the right light, a sculptural plant like this sago palm creates exceptional patterns on walls, floors, and tables. This can be achieved with either direct sun from the window or a lamp strategically positioned to spotlight the leaves.

SHEER BEAUTY

Some plants' leaves, like those of this 'Escargot' rex begonia, are even more marvelous when backlit. The sheer, speckled leaves of caladiums and the patterns of thicker marbled leaves of bromeliads and veined alocasias are also enhanced by a bit of sunshine.

TAKE THE FOCUS OFF THE TV

Create a jungle of plants around your entertainment center to keep the TV from dominating your living room. The key is to set the scene, not steal it. You want to be able to clearly see and focus on the television when it's on, so your plants should frame, but not overlap, the screen. Though there is a *lot* going on here, the backdrop is so full and the foliage and pots sufficiently unified in palette that it all blends into a pleasing wall of green. Big, colorful blooms or lots of empty spaces on the wall would fight for attention.

A CORNER FOR RETREAT

Create a special corner in your home that is all about you—a place to nourish your body and mind. What do you want to see and how do you want to feel when you're sitting there? Do you yearn for a cozy nook where you can sit quietly, beside a plant or two? Or do you crave an open and airy space to instill a sense of calm? Here are a few styles to choose from.

SLEEK

An iconic Eames lounge chair calls for an equally architectural planting, like this pencil cactus in a simple concrete pot (left). The alpine schefflera behind the chair creates an increased feeling of security and adds needed height to the display.

TRANQUIL

Everything about this pared-down look says "breathe": the soothing colors of the chair, the smooth shape of the beige vase, and the translucent, flowing curtain (which also provides filtered light). The upright stems and beautiful fan-shaped leaves of the elephant ear look as though they're ready to cool you off. *Ahhhh.*

MOD

Travel back in time with this candy-red couch, geometric pillow, and unique fish-bone cactus hung from leather straps in a repurposed bird-cage stand. This elevated display is perfect if you have limited floor space or want to create a vertical garden without making holes in your wall.

FRAME YOUR VIEW

Some windows look out on glorious rolling hills, others onto bustling streets and skyscrapers—but whatever the view, those glass panes are your home's connection to the outside world. Accessorize your outdoor scenery with a planted "frame" to bridge the gap between outdoors and in.

The alcove of this bay window is a study in how a grouping of plants can add up to more than the sum of its parts: The variegated African candelabra cactus (left) and a small fancy-leaf zonal geranium (center) make for a fine pair in their coordinating pedestal stands, but they'd look lonely if they were the only greenery in this large alcove. A modern planter (in a similar style and complementary tone; right) holding two aloe plants as well as a draping begonia and an olive plant adds some needed weight to the right side of the alcove, and the scene is finished with a hanging 'Attar of Roses' pelargonium that is just the right size for the central window. For a simpler framing device, seek out an arching potted tree. Place it so the trunk fills one side of the window and the branch curves along the top, creating a simple frame to draw the eye to the view outside.

Kitchen

The kitchen is the heart of the home. It's where people gather, cook, and nourish one another. It's also among the most convenient places to keep plants: with wipeable countertops and sweepable floors, there's a lot less worry about water damage or mess—and, of course, there's always a water source at the ready (simply soak the plants in a filled sink or give them a sprinkle with the faucet). This is a hardworking room, so in the pages that follow you'll find plants that can be put to use (including specimens that are good to eat and helpful in controlling pests) and ideas for keeping green things off the countertops and out of the way.

A COUNTERTOP HERB GARDEN

Whether you are starting with just a single plant or are a culinary master who wants a wide range of flavors at your fingertips, here are a few simple ways to grow and store herbs in your kitchen.

STAGE

To house your herbs the easy way, keep them in their grow pots and stage them in a decorative cachepot. The box shown at left fits three 4-inch (10 cm) grow pots perfectly and is outfitted with a tray to catch any extra water. The blue bowl is home to a strawberry plant. Pop pots in and out as the seasons change—or as culinary inspiration strikes!

HANG

Repurpose a multitiered fruit basket into your own hanging herb garden. Pair with inexpensive terra-cotta pots, painted in the colors of your choosing. Pictured here is a mix of mint, thyme, and sage (top tier) and basil and oregano (bottom tier). 'Kent Beauty' oregano is the perfect hanging plant, as it offers up pendulous flowers with a profuse fragrance.

CUT

After buying fresh-cut herbs at the market, give the bottoms of the stems a fresh snip, fill a vase with cool water, and place the herbs inside. Remove any leaves that are underwater to help keep the water clear. Although this storage method is temporary, the herbs will fill your kitchen with their alluring fragrances while they last!

DRY

Add a colorful display to your kitchen wall by hanging herbs upside down to dry. After a few weeks, store in an airtight container out of direct light and use within a few months. If you'd like to keep the herbs on display, oregano, sage, and thyme dry more beautifully than rosemary and basil. The "drying rack" shown here is actually a simple dish rack.

CLEAR YOUR COUNTERTOPS

If space is tight, or you just want to add a bit of interest to your walls, here are two ways to elevate your potted plants. They will be heavy, especially after they've been watered, so make sure that whatever hanging style you select, you drill into a stud or use a wall anchor.

WALL HOOK

This vintage "woven" metal cachepot is the perfect foil to the large, simple shapes of wax plant (*Hoya obovata*) planted in a simple terracotta pot. The pretty copper hook it's hanging from is from Terrain, but you could easily paint a generic hardware-store hook whatever tone you'd like. After installing your hook, measure the radius of the pot you'd like to hang (the length from the center to one of its edges) and compare that to the distance from the end of the hook to the wall to be sure there is enough clearance. Care tip: If you're often sautéing on your stovetop, your plants may soon bear a layer of kitchen grime. Keep them clean by gently wiping them occasionally with a damp, soft rag.

LOOP HANGER

This iron loop hanger (available in hardware stores) conjures up images of brightly colored pelargoniums hanging en masse on the side of an old French farmhouse. To replicate the rustic vibe, this Swedish ivy (*Plectranthus ciliatus*) was kept in its original grow pot and placed into the terra-cotta container to prevent dirty water from dripping onto the counter after watering.

PEST CONTROL

While most plants acquire nutrients through their roots, carnivorous plants have developed techniques to lure insects into their "pitchers," trap them with their sticky pads, and suck them down their slippery leaves. As the insects decay, the plant soaks up all their nutrients. Sounds like a science project, huh? The Venus flytrap is certainly the most famous carnivorous plant, but there are other bizarre and beautiful specimens that make for prehistoric-looking, practical displays as well.

MONKEY CUPS

Monkey cups (*Nepenthes* spp.) are naturally found in tropical rain-forest canopies, where they catch water and bugs from the sky (and where monkeys can enjoy a drink from their pitchers—thus their funny name). These plants love humidity and should be kept damp at all times. Unlike their sun-loving carnivorous companion the pitcher plant (see opposite), they do not like direct sun.

PITCHER PLANT ⌃

Pitcher plants (*Sarracenia* spp.; left) can be found in both short varieties that grow to just a few inches (7.5 cm) high and taller ones that can reach up to 30 inches (76 cm). Pair your pest controller with another hardworking kitchen favorite, an Arabian coffee plant (*Coffea arabica*; right). With patience and a whole lot of light, these evergreen plants will produce raw coffee beans (in three to five years).

SUNDEW

The sundew (*Drosera* spp.) is a bog plant and prefers a wet environment. Here pebbles and moss were added to polish off the top attractively. Place the sundew in a bright place and keep it moist, and this wondrous kitchen companion will lure fruit flies and other pests and wow your family and friends.

STELLAR SILLS

Even the narrowest of windowsills can hold a few plants and bring life to your kitchen. Remember, though, that this area is often susceptible to the outside environment—winters can be cold and drafty, and summers can get quite hot. Move your friends out of these extremes as needed.

LOW-LIGHT STUNNERS

This lineup of tried-and-true low-light plants gets extra zing from an eclectic mix of containers. The tallest specimens are housed in mugs of roughly matching heights (at far left, a ZZ plant; center, a spear sansevieria). A pair of short ceramic saltcellars hold a 'Red Ripple' peperomia (second from left) and haworthia (far right)—the plants themselves were placed at a slant to match the containers' profiles. The only "real" pot here is a teeny-tiny vessel holding a teeny-tiny ivy (second from right).

MIX-AND-MATCH SUNNY SUCCULENTS »

A brightly lit windowsill calls for a wild menagerie of succulents. If space is at a premium, gather your favorites into a single mug—pictured here are watch chain, variegated corncob cactus, 'Lola' echeveria, thimble cactus, and Siebold's stonecrop. The plants' muted tones mean the variety of shapes and textures is the focus. Keep your collection growing by plucking any offshoots—simply set them on some soil to root and watch them thrive.

TABLETOP JEWEL BOX

If you have a smallish table, or one that is frequently commandeered for not only the dinner rush but also rainy-day art projects and late-night homework sessions, you may not have a lot of space for additional decor. But even a *touch* of green will liven up your space. Just choose a planting that takes up a small footprint, with a low pot that is tumbleproof! This grouping of exquisite yet easy-care plants combines an angular 'Black Mystic' earth star with the fluffy flowers of the cooperi crassula and a bright 'Ruby Blush' echeveria for a dynamic push-pull of color and texture. The solid greens of the emerald ripple peperomia, 'ihi, and soft gray echeveria serve as a unifying backdrop.

OPPOSITE: *The bowl's gold hue connects the centerpiece to the other decor in this dining nook, including the wall mirror, light fixture, and pillows.*

ABOVE IT ALL

For an unobtrusive yet striking centerpiece, consider the space over your head! The designs below pair plants with a hanging lighting fixture, providing cozy ambience around the table.

PLANTED PENDANT

Haworthia, a few different species of rhipsalis, rat's tail cactus, and a pickle plant make perfect companions in this high-style light fixture–as-planter by Object/Interface. Since the bulb lights the table below, not the plants above, make sure these sun lovers get some natural light.

HANGING TERRARIUM »

For another double-duty design, check out this light terrarium by Lightovo. A sweet hanging home for this moody, wild grouping, the fixture lights both the table below *and* its plant occupants. Here a rex begonia, a Sprengeri fern, and arrowhead plants were wrapped together in moss, kokedama-style (see page 102), then inserted moss-ball first into the terrarium. A single plant in a tiny pot would look nice, too. Gently water with a turkey baster or watering can with a narrow spout to avoid a pool of water in the bottom of the glass.

Bedroom

We spend a third of our lives in bed, so our sleeping spaces should be as appealing and comforting as possible. This chapter focuses on the ways plants can help us achieve this. Amazingly, just one plant can make your bedroom air a bit fresher. You'll also discover plants that help mitigate noise and others that respond to light in a surprising and delightful way. When you're in bed, it's lovely to look out on a beautiful array of plants and connect with the calming and restorative effects of nature.

A JUNGLE HIDEAWAY

This cozy bedroom features more than a dozen plants in a pleasing assortment of forms and textures: airy, upright, sprawling, glossy. The secret to designing a full space like this one, which feels comfortable and intimate rather than crowded and claustrophobic, is to create "zones" within the room and make sure each grouping has a mix of sizes and shapes to mimic what happens in nature. Here are a few more tips.

TAKE IT ONE STEP AT A TIME. Let your collection grow slowly. A nightstand topped with an assemblage of plants will offer a wilder, fuller feel than the same few plants scattered around the room.

AIM HIGH. In a real jungle, plants appear at every level, including above your head. So add a floor plant that is reaching for the ceiling (and let it drape slightly, like a canopy over the bed). You can create a similar effect with hanging planters: hang several at different heights to create a living canopy.

ADD A FOCAL POINT. Keep the design interesting yet restful on the eye by opting mostly for simple, soft shapes in a harmonious green color theme. But be sure to incorporate a couple of plants that stand out from the crowd due to their eye-catching color or shape. Thanks to its unusual curved trunk and huge odd-shaped leaves, the snowflake aralia (right) really holds its own, even in a busy crowd.

CLEAR THE WAY. Once you get bitten by the plant bug, it might be hard to rein yourself in. But remember to keep a path wide enough to negotiate comfortably. This will protect those delicate leafy friends and save you the struggle of navigating an obstacle course each time you cross the room.

A MINIMALIST OASIS

Thanks to the exceptional large bay window, little is needed in this bedroom in the way of decoration. For a clean and calm look, you can't beat a simple white-and-green palette. Though the vining 'Jade' pothos plant below adds a bit of color and interest, casting a green hue as sunlight filters through its leaves, consider the planting options opposite if you're in search of some privacy.

SCREEN FROM ABOVE

To create living curtains, insert a row of hooks along the ceiling and hang plants from them (for a utilitarian look, simply keep your plants in the plastic hanging grow pots they are sold in), or install a shelf just above the window and let full, draping specimens like this creeping ficus hang from it.

SCREEN FROM BELOW

Add privacy from below by lining up a row of tall specimens like these wide-leaved snake plants. If they need a boost, use a box, books, or a plant stand.

NIGHT MOVES

Believe it or not, some plants turn in for the night, too. This sleeping movement, called nyctinasty, occurs in response to darkness. Once the light levels drop, the leaves close up for the night. As the sun rises, you will see the plant "stir" from its slumbering state as its leaves stretch out to collect energy, making it a lively addition to a sunny room. The sensitive plant pictured here closes at night and also reacts to the slightest touch (a response called seismonasty, a defense against predators), which is why it's also commonly referred to as the touch-me-not plant. It's a slightly fussy houseguest, but when treated right, it is a fast grower; prune it back to keep its mounding habit (as shown opposite) or it will become tall and lanky. Keep it near your bedside, and you'll enjoy the rhythm of the earth's rotation together.

LEFT: *This sensitive plant's proper name,* Mimosa pudica, *aptly warns of its collapse when touched (or in response to darkness):* pudica *is Latin for bashful. Once given peace and quiet and some sunlight, it gently opens its leaves again.*

ONE DRESSER, FOUR WAYS

Turn your dresser into a vignette that celebrates your personal style. Choose easy-care plants like the ones featured here to create ambience without adding to your busy schedule.

URBAN BOHEMIAN

Introduce a laid-back feel by softening bold features, like this large mirror, with a vine or other climbing greenery. Here a grape ivy was attached to the mirror's frame using clear 3M hooks (which are discreet and removable).

FEMININE GLAMOUR

Combine softly textural plants—like the delicate beading of string of pearls (above) and the voluminous tousles of 'N'Joy' golden pothos (below)—with statement-making metallic accents to create an elegant space perfect for pampering and indulgence.

SLEEK CONTEMPORARY

Monochromatic colors and subtle patterns create layers of interest without being overwhelming. Here the sophisticated metallic vase mimics the shape of the mirror frame, and the unusual, architectural 'Ming Thing' cactus within sparks interest.

NATURAL BEAUTY

The vibe here is Scandinavia meets Joshua Tree. Both the plant and the accessories are monotone, so varying textures are key to adding interest: the woven basket against the painted ceramic planter and the soft furry leaves of the 'Chocolate Soldier' panda plant.

TALL, DARK, AND HANDSOME

Trick your eye and create a space that feels big on green but utilizes plant design with a small footprint. If you're tight on space, these strategies will help you squeeze in more greenery without taking up valuable real estate.

PACK IT IN

To add a whole lot of plants into a narrow space, fill a slim-line vase with a smorgasbord of sansevierias. The standard snake plant is included here, as well as interesting varieties with fabulous names such as shark's fin, bird's nest, starfish, and 'Superclone'!

TRAIN IT

Variegated fatshedera (pictured here) and other climbing vines can get unruly, but tightly prune and secure the vines to a trellis with twist ties or twine (a technique called espaliering) to keep them contained and almost flat against the surface.

GIVE IT A BOOST

Make a small or medium plant pull the weight of a floor plant, but in half the space, with a plant stand. This one is only 7 inches (18 cm) wide and snugly fits a plant's vase (or one can be set on top with the addition of a plate). Go even taller by choosing a vertical specimen like a snake plant or, if you have a smidgen more room to spare, add a draping plant like this rex begonia vine.

Bathroom

Yes, a bathroom is a necessity, but it can also be a place for relaxation, comfort, and pleasure (bubble bath!). Plants are wonderful enhancements in this space—and luckily, bathrooms are welcoming places for greenery. They tend to have decent heat and humidity, which some plants require, and, like the kitchen, bathrooms typically offer water- and stain-resistant surfaces. Plus, it's the easiest room in your home in which to water your plants. Use the showerhead to mimic rainfall to clean leaves and soak the soil—but place a screen over the drain to catch dirt, gravel, or whatnot so that it doesn't clog the pipes.

SOAK UP THE SUN

A sunny corner of the bathroom is freshened by a palette of minty green. This arrangement is a lesson in the importance of layering for scale and interest. If all of these plants were at the same height as the tub, it would make for a one-note display, and the tub would overpower the space. Instead, each plant is on its own level. The hanging succulent string of bananas draws the eye upward, but its slim, tidy silhouette keeps it from dominating the view. The naked feltleaf kalanchoe was placed in a footed planter and further elevated on a stool, and the wheeled planter features a tall pencil cactus and the lower, denser foliage of silver plectranthus. Even the tiny nutmeg-scented 'Old Spice' pelargonium is given a boost by an overturned woven basket.

Thanks to this layered approach, the tub doesn't feel outsize, and the whole vignette is visually appealing. This eclectic collection likes bright light, so open the window and let the sun shine in.

Plants (like the ones shown here) with hints of teal and gray-green hues, as well as leaves that are smaller, narrower, or more complex in shape, have evolved to tolerate bright and direct light. In contrast, darker shades of green and larger leaf structures (like those shown on page 100) suggest that a plant has adapted to low-light conditions.

A LOW-LIGHT "TUBSCAPE"

When filling a space with loads of plants, think of your room as an indoor jungle. In the wild, some plants reach to the light of the sky, while midsize specimens thrive under the shade of a tree. Still other plants grow low to the ground, drinking up the dappled light. So when you approach your room design, consider these three planes: the space overhead, at eye level, and on the ground. If you include plants on all three levels, the space will feel more like a natural environment—you'll have to look up, down, and around to catch it all! Here's the breakdown in this space.

OVERHEAD: Hanging plants like the spider plant (upper right) and tall floor plants like the Madagascar dragon tree (far left) not only envelop the scene in a canopy of green but also provide dramatic scale.

EYE LEVEL: Eye level is our most common vantage point, and plants in this zone will deliver instant impact whenever we walk into a room. The snake plant (center) and monstera (right) work particularly well at this level because their bold-shaped leaves provide interest when viewed from the side.

GROUND LEVEL: Plants placed on the ground or at lower levels will frequently be seen from above. The bird's-nest sansevieria shown in the gold vase here is fairly plain when viewed from the side, but glance down while entering the bathtub and you'll appreciate its charming rosette.

TINY TOUCHES FOR THE BATH

Use tiny utilitarian containers in unexpected ways to turn pint-size spaces into impressive plant displays.

AT THE SINK

Kokedama, a Japanese art form that translates as "moss balls," comes in many forms. This 'Jeanette' dwarf English ivy kokedama is perfectly suited to the bathroom sink—set on a soap dish, it's a snap to care for: simply soak in the filled sink for twenty minutes, remove, gently squeeze out excess water, and return the ball to the soap dish to finish draining. Regular misting will keep the moss looking vibrant, too.

IN A SOAP DISH

Corner soap dishes that come with suction-cup backs make perfect niches for air plants like these *Tillandsia stricta* 'Houston Dark Pink' and Spanish moss. After a shower, be sure to shake off excess water so that it doesn't pool in the dish and rot the moss.

ON THE MIRROR

This tiny suction-cup holder meant to corral shaving razors (a hardware store find!) works perfectly for plant cuttings. Tropical vine-like plants such as philodendrons and pothos take well to rooting in water—just change the water occasionally and watch these babies grow. When the roots get too long, plant the cuttings in soil and replace with some fresh cuts. See page 34 for more on propagating plants. On the counter below, a small cup (part of a vanity accessories set) was repurposed as a pot for a little succulent.

FROM THE TOWEL RACK

This window box is meant to hang on an out-door railing, but why not hang it on your towel rack instead? Here it is filled with four types of bromeliads (from left to right): earth star, guz-mania, vriesea, and 'Hallelujah'. Bromeliads drink from their rosettes, so water them from above, allowing a tiny pool to gather in each cup. Be sure your towel rack is mounted securely and can take the weight of the plants, then set the grow pots inside (eliminating the extra weight of soil packed between plants).

A GREEN APOTHECARY

Medicinal herbs have been used for centuries, by ancient Egyptians and Babylonians, medieval monks, and even modern-day physicians. Many of the findings from early herbalists became milestones in Western medicine as we're familiar with it today. There's something particularly rewarding about growing plants with a practical use, especially when they provide you with not only a stylish medicine cabinet but also the comfort of knowing a homegrown first-aid kit is always on hand. Whether you're going to use the scent to relax or the gel to soothe, all of these plants are helpful additions to a sunny bathroom.

1. **ALOE:** This succulent (*Aloe vera*) is durable, easy to grow, and quick to multiply. When the leaves are cut, they release a gel that soothes burns and promotes healing — and can be added to homemade beauty products like face washes and hydrating masks.

2. **ROSEMARY:** This herb has uses that extend far beyond culinary creations. Cut off a piece or two, put in hot water, and breathe in the scent for a quick pick-me-up and to help open up a stuffed nose during allergy or cold season. Rosemary (*Rosmarinus officinalis*) comes in both upright and draping varieties. To re-create this bonsai-style form, pick out a draping variety in a 4-inch (10 cm) grow pot and prune to shape.

3. **COCONUT PALM:** If planted in the ground outside in a warm climate, this palm (*Cocos nucifera*) will grant you coconuts in about five years (it will also grow up to 50 feet/15 m tall). You can bring it indoors and plant it in a small pot to keep it manageably small, but then you're not likely to see fruit. Even if you can't drink the milk and reap the medicinal benefits, it should soothe the soul with its tropical innuendoes.

4. **LAVENDER:** This workhorse (*Lavandula* spp.) acts as an air purifier, and its aroma reduces anxiety. Release its fragrance by simply rubbing the flowers or leaves. Lavender dries extremely well, too — collect the petals into a small sachet to tuck under your pillow or place in your drawers to subtly scent your clothes or sheets.

A LIVING SHOWER WALL

A sunlit shower is the perfect host to a collection of mounted epiphytes—each plant gets sprinkled with water and can enjoy high humidity and bright light. Ready-made mounted plants are available from garden centers and online crafters, but they're easy to make on your own. Simply follow the steps below.

WHAT YOU'LL NEED:

1 epiphytic plant (a plant that grows on other plants, rather than in soil); pictured here are a Jenkins's dendrobium orchid (top) and a *Tillandsia brachycaulos* × *concolor* air plant (center)

Cork bark

Fishing wire

Moss or colored twine (optional)

Waterproof, heavy-weight-bearing adhesive hooks

1. Rest the plant horizontally on a piece of cork and wrap fishing line around the two to secure.

2. Add moss or colored twine if you wish, for further decoration.

3. Attach to a bathroom wall with adhesive hooks.

Below the mounted dendrobium orchid and air plant is a handmade wall-hanging orchid pot (opposite). This one has a flat back and a carved design that allow for easy hanging, air circulation, and drainage, perfect for an upright miniature orchid and dangling pink lipstick plant (Aeschynanthus 'Thai Pink').

Home Office

The benefits of green things is a topic that's hot in the press and backed by science, too. And now, researchers from the United Kingdom, the Netherlands, and Australia suggest that we make fewer mistakes, have more positive feelings, and can be more productive when surrounded by even a few houseplants in the office. After studying the subject for a decade, the team concluded that folks working in an office landscaped with plants can see a 15 percent increase in productivity compared to those working in stark office environments. The following pages offer plenty of efficiency-boosting specimens. And because a room often filled with paper and expensive technology may not be conducive to lots of potting soil and water, there are plenty of mess-free options here as well.

TO THE MAX

For a desk that's bursting with inspiration, create a lush display filled with plants of all sizes, as well as tchotchkes, memorabilia, favorite books, and quirky vases. The central shelves of this flexible shelving system were aligned to help "center" and lend a sense of order to this busy design. For interest, though, the top shelves were staggered—and the highest shelf is much deeper than those below it. Books were used to add height (and serve as mini plant stands) and were stacked both horizontally and vertically to help break up the lines of the design.

If you're creating a shelving unit with a bounty of plants, be sure to add some with distinctive personalities, like the twirls of 'Frizzle Sizzle' albuca, tiny fruits on the mistletoe fig, or the curved trunk of a bonsai ficus. Draping plants are great for breaking up the lines in the shelves; try the kangaroo ivy, glory bower, or prayer plant. Then fill in with solid green foliage with some visual interest, like the coin-shaped ones of the *Pilea peperomioides* and the ear-shaped leaves of the large 'Congo' philodendron.

RIGHT: *This colorful foliage from a ti plant was originally meant for a cut floral arrangement. Give it a new life as a houseplant by snipping off a few stems and placing them in water in a somewhat bright area. Transfer to a potting mix after roots form and place in an area of medium to bright light. Allow the soil to dry out between waterings.*

CONTROLLED NATURE

Are you attracted to the tidiness of a freshly mowed lawn? Does a precisely trimmed topiary woo you? If so, you're attracted to controlled plant design rather than wild and woolly schemes. Choose a few clean-lined plants like a braided spear sansevieria (pictured here on the desktop) to convey control and stability and introduce natural repetition and muted colors into your setup to help make for a calm, orderly space. Here are a few tips.

STAY ON THE SAME PLANE. All the shelves in this desk unit are aligned, so the foundation of the design is planned and predictable.

GROUND WITH COLOR. The monochromatic gray-blue palette evokes the feeling of water and the sky, inducing a sense of calm.

FOLLOW THE RULE OF THREES. Engaging, effective, and efficient, things presented in threes are pleasing to our eyes. Here the three pots on the top row allow the human brain to create a center point and rest the eye as the two others add balance. Choose three vases and plant combinations that are similar (here they all are planted with pinstripe and rose-painted calatheas; the center one varies slightly, with an added bird's-nest sansevieria).

CREATE A PATTERN. Repetition ties the look together and generally makes us feel secure. Here the connecting threads are the metallic objects and the "natural" accents. The rock in glass is an art piece by Lawrence LaBianca, the eggshell-embellished vases evoke robins' eggs, and the large pot on the floor (holding a Japanese aralia) looks like it's made of stone.

KEEP IT SIMPLE. Here's a novel, hands-off way to bring something intriguing into the office space: the miniature orchid on top of the file boxes here, sold as *Psygmorchis pusilla* or *Erycina pusilla*, stands just 2 inches (5 cm) tall in its enclosed environment and produces yellow blooms.

TINY DESKTOP TOUCHES

Not only are air plants mess-free, they also can be stuck in just about any out-of-the-way spot on your desktop! Below are a few ways to display them.

WIRE IT

Wrap a heavy-gauge but malleable wire around the base of your air plant (*Tillandsia brachycaulos* × *abdita* is shown here) and insert it into a tiny drilled hole in a block of wood to create a miniature sculpture. Remove the plant from the block to soak, or mist it.

GLUE IT

Use a glue gun or nontoxic clear, waterproof glue to attach your specimen to a paperweight, a magnet, or a little vase like this one. If using hot glue, let it dry for a moment before applying the plant so you don't burn it. Hold the air plant in place for a bit to stabilize it, then let the glue set before hanging.

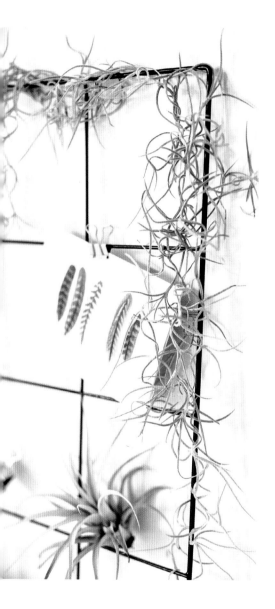

TUCK IT

Spanish moss or even small air plant varieties are tiny enough to tuck into a wire organizer (as shown here); you can also clip the plants to the wire with a binder clip. They are easy to remove, water, and replace.

FOUR MINIATURE LANDSCAPES

If your space doesn't have a window, build a little plant scene and daydream of the great outdoors. Give yourself a desktop reminder of that beach vacation from last year, or re-create a calming walk in the wilderness. What escape will you create?

A SUNNY DESERT

This small ox tongue succulent (*Gasteria limpopo*) rests in what was intended as an ashtray but now serves as the planter's coaster. Place the succulent in the bowl of the ashtray, fill the opening with cactus mix, and display in a moderate to well-lit spot. Add a thin layer of stones on top of the soil to complete the handsome display.

A LAKEFRONT

Awaken the memories of summer with a tiny water garden made of rocks, sand, and marimo. Keep it cool and in low light (no direct or bright light, please). Flush the old water with clean, cold H_2O every ten days or so. As an added benefit, according to feng shui tradition, still water is said to improve wisdom, clarity, and wealth.

A FOREST FLOOR

This terrarium-esque vessel by Modern Sprout is made specifically for moss, but you don't need to purchase a special display: simply plop moss in a low bowl and you can almost feel the softness of green underfoot. The lid keeps the moss moist, but if its color fades from bright green to brownish, it needs additional humidity: mist or remove and give it a soak. Keep out of bright or direct light.

A SANDY BEACH

Choose any clear glass vessel, pour in two tones of sand one after the other, add a wave design by puncturing one layer into another with a toothpick, and top with a tiny air plant. Every few weeks, remove the air plant and soak it, then shake off excess water, pat dry, and replace.

GREEN DESK ORGANIZERS

If your desktop real estate is particularly precious, consider wall-mounted vessels. And while elsewhere you've seen how to repurpose household items as containers for your plants, here we do the reverse—leave a couple of planters empty and use them as tidy homes for office supplies!

GRAPHIC DESIGN

These narrow peach vases keep a low profile on your wall. If you have a dark office, use them to hold ivy (left) and a pair of spear sansevieria (right)—plants that won't take up much space and can handle low light levels and a bit of neglect. Be mindful of ivy: if you don't want it to climb and stick to your wall (and potentially cause damage to your paint), choose a different green companion.

INDUSTRIAL CHIC »

Hang these leather-and-rope planters from hardware accessories-turned-hooks for a rugged, DIY look. The connection between these pencils and their plant companion extends beyond their yellowish hue—this sculptural specimen is commonly called a 'Sticks on Fire' pencil cactus. This vibrant plant keeps its orange coloring best in direct light.

A POP OF COLOR

These four weird and wonderful desktop plants bring a figurative breath of fresh air to your office space, especially when paired with equally vibrant vessels!

CROTON

The leaves of croton plants come in varied colors, from yellow and green to orange. Their patterns are equally diverse—they can be spotted, veined, striped, or speckled. These eclectic attributes are the reason for the plant's other common name, Joseph's coat (in reference to the many-hued coat of biblical lore). Embrace its bold nature by giving it an equally color-saturated pot. Be warned that this houseplant is rather fussy—it likes bright light indoors, high humidity, and soil kept slightly moist.

EARTH STAR

This 'Pink Starlite' earth star (also called a bromeliad) will be at its most colorful if grown in bright light. It benefits from humidity, so mist often (it's a perfect fit for tidy desktop terrariums, too!). Contrast its beautiful pink leaves with an underplanting of moss. Let it tip over the side of a low vase to give yourself a lovely view while you're hard at work.

CROWN OF THORNS

Paired with a handsome blue vase, this exquisite bloomer is lovely to look at, though not to touch. Don't let the pretty yellow flowers of this cultivar (*Euphorbia milii* 'Dinni Yellow') fool you—this plant is aptly named, bearing sharp thorns underneath its succulent foliage. More commonly found with red or orange flowers, crown of thorns plants need direct light and dry soil. In some cultures, they are believed to confer good luck on the plant owners and their homes.

NEON CACTUS

This is actually two plants in one. The top cactus (*Gymnocalycium mihanovichii* var. *friedrichii* 'Hibotan') has a mutation that causes it to lack the chlorophyll that makes plants green, thereby exposing the red, orange, or yellow color beneath; it is grafted onto another, taller cactus (usually *Hylocereus undatus*). This cupcake-liner-like vase makes the pairing extra sweet. Give it bright light, and let it dry out between waterings.

With so many great retailers and artisans out there and more popping up all the time, it's impossible to include everyone here—but below are some of the best and most easily accessible sources for plants, supplies, vessels, and more (and the places where much of the product for this book came from). Be sure to look out for local plant swaps and gatherings, too, and check out plant societies, botanical gardens, and educational centers in your area—they are filled with excellent information and inspiration.

PLANTS & SUPPLIES

BLACK JUNGLE EXOTICS

blackjungleterrariumsupply.com

This supplier offers a wild selection of plants, including carnivorous specimens, as well as LED grow lights.

BRENT AND BECKY'S BULBS

brentandbeckysbulbs.com

Check out this site's indoor bulbs section.

COSTA FARMS

costafarms.com

This company is a wholesaler, but you can learn a lot from their houseplant-focused blog.

GARDENER'S SUPPLY COMPANY

gardeners.com

A one-stop shop for all your gardening needs.

KARTUZ GREENHOUSES

kartuz.com

The go-to spot for all your begonia needs!

LOGEE'S

logees.com

This company specializes in fruiting, rare, and tropical plants.

PISTILS NURSERY

pistilsnursery.com

Look to Pistils for hip plants and helpful info.

THE SILL

thesill.com

They deliver houseplants directly to your door, anywhere in the United States.

VESSELS & PLANT STANDS

ARTISANS

There is a whole world of talented artisans to explore, including the makers of some of the handcrafted vessels in the book. Search for them online to find out how to get your hands on a piece of their art: Aveva Design, Ecoforms, Eric Trine, Esther Pottery, FreeFolding, Global Eye Art Collective, Heath Ceramics, Holly Coley, House of Thol, Judy Jackson, Kelly Lamb, Love Fest Fibers, Melanie Abrantes Designs, Michiko Shimada, Modernica, Paige Russell, Pawena, Pseudo Studio, SaraPaloma, West Perro, and Yonder SF.

CAMPO DE' FIORI

campodefiori.com

When you want to bring an old-world garden feel to the indoors, you can"t beat Campo's mossy terra-cotta pots, wrought-iron plant stands, and handblown glass terrariums.

ETSY.COM

This online marketplace for makers is a great place to find unique pottery and plant accessories.

FERM LIVING

fermliving.com

A mecca for all things Danish design, including a fantastic collection of minimalist pots and plant stands.

FLYING TIGER COPENHAGEN

flyingtiger.com

Check out this playful store's ever-rotating collection of lighthearted and colorful pieces, at prices that won't empty your wallet.

KAUFMANN MERCANTILE

kaufmann-mercantile.com

For everything from garden tools to aprons.

MODERN SPROUT

modsprout.com

Modern Sprout's delightful goodies include attractive grow lights, mod terrariums, and even herb kits.

POTTED

pottedstore.com

Whether you're looking for a succulent wall planter or a colorful hanging vase, this LA-based store has got you covered.

TERRAIN

shopterrain.com

Terrain is synonymous with well-made products that bridge the gap between indoors and out.

ACKNOWLEDGMENTS

I have such gratitude for the talent, knowledge, and creativity of the whole team involved with the making of this book: First, my editors, Bridget Monroe Itkin and Elise Ramsbottom, for their guidance, kindness, and expertise down the bookmaking path. Aubrie Pick for having the amazing combination of a down-to-earth demeanor and over-the-top photography skills. Sarah Green, whose behind-the-scenes talent shines throughout the book. My agent, Kitty Cowles, for always steering me in the right direction. Alexis Mersel, for helping get this idea off the ground with eloquence. Molly Watson, for once again transforming my mishmash of words into a cohesive narrative. Lia Ronnen for having the faith to work with me on another project. The rest of the Artisan team, including Carson Lombardi, Paula Brisco, Barbara Peragine, Nina Simoneaux, Michelle Ishay-Cohen, Jane Treuhaft, Nancy Murray, Allison McGeehon, Theresa Collier, and Amy Michelson, for all the help along the way. Erin Heimstra and Kate Leonard, for their keen eyes and gift for style. Cortney Munna for managing the shoots and postproduction with clarity and enthusiasm. Tony Colella, Kana Copeland, Bessma Khalaf, and Sherese Elsey, cheers to the finest plant and photo assistants ever. A nod to Sophie de Lignerolles, a talented artist, gardener, and dear friend. And to Shannon Lynn, who pitched in and helped lead Lila B. when an overlapping project needed a pilot.

Thanks also to my friends, colleagues, and neighbors who generously fielded my questions, opened their homes, and lent out their belongings—they are a huge part of what's between these book bindings. To jak.w, who elevated this project when they opened their fabulous studio full of treasures to us. Thank you to Erin Heimstra, Chris Wick, and Benni Amada for allowing us to move into your picturesque homes for a few days. To all my neighbors in the Allied Box Factory for putting up with all my green friends that overflowed into our communal hallways and for pitching in when I was searching for the perfect prop. To Granville Greene, Jackie Priestley, and Lawrence Lee for solid advice in your fields of writing, coaching, and horticulture. I feel lucky to have you all in my life! To the folks at Delano Nursery (John, Lauren, Carla, Penny) for not only offering cool plants within the San Francisco Flower Mart but also for loaning me "Phil," your precious 'Jade' pothos. And to Rocket Farms, Costa Farms, Hana Bay Nursery, and the San Francisco Botanical Garden for giving me sound horticultural info. To the talented artists whose artwork dons the walls in this book: Sidnea D'Amico (page 94), John Fraser (page 94), Ian Green (page 59), Lawrence LaBianca (page 117), Rex Ray (page 116), Melinda Stickney-Gibson (page 53), and Myrna Tatar (page 8). And finally, a whole bunch of gratitude goes to my sister, Suzanne, who graciously carried the load of our shared endeavors while I concentrated on the book.

INDEX